Outdoor Electrical SAFETY CHECK

TIPS FOR THE SAFE OUTDOOR USE OF ELECTRICITY

ESFi
Electrical Safety Foundation International

Published as a public service by the Electrical Safety Foundation International in cooperation with the U.S. Consumer Product Safety Commission and the Canada Safety Council.

CONTENTS

Introduction 2

Electrical Safety Devices 3

    What are electrical safety devices? 3

    Ground fault circuit interrupter (GFCI) 3

    Arc fault circuit interrupter (AFCI) 4

    How do electrical safety devices work? 5

    Hazards in the use of electrical
        products outdoors 12

    Before using electrical products outdoors 14

Safety Rules 17

    Hot Tubs, Spas and Pools 20

    Extension Cords 21

    Electrical Lawn and Garden Products 23

    Battery Operated Products 27

    Power Tool Safety 29

    Power Line Safety 32

    Electrical Safety During Storms 33

    Lightning Safety 36

Glossary 38

Note: Throughout the pamphlet words in blue are listed in the glossary.

INTRODUCTION

Most of us think we know enough about electricity to stay safe. We don't stand under trees during a thunderstorm. We don't go near downed power lines. We use portable GFCIs and follow manufacturer's instructions when using portable generators. We make a habit of electrical safety.

But each year, hundreds of people die and thousands of people are injured because of electrical hazards, nearly all of which could be avoided through good electrical safety practices.

ESFI's *Outdoor Electrical Safety Check* is a great reference tool for electrical safety. It explains how you can use electrical safety devices to help protect against conditions that can cause electrical shock and fire hazards. It includes safety tips for dealing with power tools, downed power lines, floods and electricity, and safe use of generators.

Whether at home or in the work place, this checklist can help you identify and address electrical hazards. Use this as a reference and follow these guidelines to make your outdoor life safer.

For more electrical safety information, visit our web site—www.electrical-safety.org!

ELECTRICAL SAFETY DEVICES

WHAT ARE ELECTRICAL SAFETY DEVICES?

Four devices that help provide outdoor electrical safety:

Circuit breakers or fuses protect against overcurrent conditions that could result in potential fire and shock hazards.

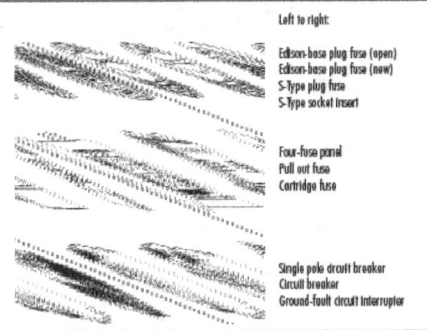

Left to right:

Edison-base plug fuse (open)
Edison-base plug fuse (new)
S-Type plug fuse
S-Type socket insert

Four-fuse panel
Pull out fuse
Cartridge fuse

Single pole circuit breaker
Circuit breaker
Ground-fault circuit interrupter

Ground-fault circuit interrupters (GFCIs) protect against potentially lethal shock when they detect even minute, but potentially dangerous ground faults, or "leaks" of electrical current from the circuit. GFCIs may be incorporated into circuit breakers protecting the entire circuit, outlets protecting everything on the circuit downstream from the GFCI outlet, or as portable devices that can be used at an outlet to give protection for a particular electrical item.

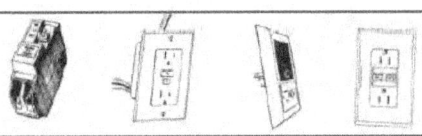

Three-pronged plugs and outlets, and polarized plugs and outlets offer enhanced protection against potential shock when provided on specific products. These measures should never be circumvented by sawing or breaking off the third prong or attempting to widen an outlet slot.

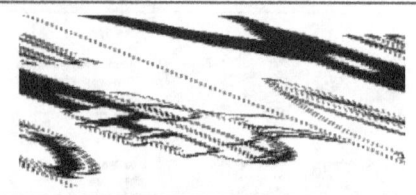

Arc fault circuit interrupters (AFCIs) are relatively new devices that protect against fires caused by the effects of unwanted electrical arcing in wiring. An AFCI will de-energize the circuit when an arc fault is detected.

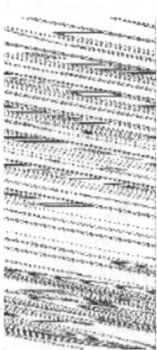

Electrical wiring in buildings with areas exposed to the outdoors, including circuits in garages, porches, patios and storage areas, could benefit from the additional electrical fire prevention features of AFCI devices when incorporated in the branch circuitry.

HOW DO ELECTRICAL SAFETY DEVICES WORK?

Circuit breakers or fuses in your home electrical panel sense overcurrent conditions and short circuits and reduce the risk of fire in your electric wiring. When you overload a branch circuit by plugging in too many products, the fuse blows or the circuit breaker trips to shut off power.

Up-to-date single-family dwellings should be provided with at least one branch circuit that carries power to an outdoor outlet. Locate your outdoor branch circuit(s) on the listing of branch circuits on your electrical panel. (If you have no outdoor wall outlet, call a qualified electrician to install one.) You should find the amperage on the circuit breaker or the fuse.

To figure out whether a combination of products will overload a branch circuit, add up the power ratings (watts) you plan to use at the same time on that circuit. The power (watts) or amperage of an electrical product is shown on its attached nameplate.

> Volts (also on nameplate) x Amps = Power (wattage). For example: 120 V x 15A = 1800 W. Demanding more than 1800 W will overload a 15 ampere circuit.

Outdoor electrical products that may use a significant portion of the power a branch circuit can supply are electric lawn mowers, leaf blowers and snow blowers.

Be sure to figure total wattage in advance when you are planning an outdoor event. Add up the power ratings of everything you will use: garden lights, electric grill, hot tub and so on plus everything else on the circuit. If you exceed the circuit wattage limitation, you will likely trip a circuit breaker or blow a fuse which can cause hidden damage to the circuit. If necessary, plan to redistribute your power needs to more than one branch circuit, or reduce the electrical load to avoid the overload situation.

A short circuit in a product, cord or plug may also trip your circuit breaker or blow a fuse. If you can identify the product that is causing the problem, take it to a manufacturer-recommended repair facility. If you don't know what is causing your circuit breaker to trip or fuses to blow, call a qualified electrician.

A ground-fault circuit interrupter (GFCI) will disconnect power automatically when a plugged-in electrical product leaks electricity to ground. Outdoors, where water and electricity can easily inadvertently come together, a GFCI is a lifesaver, not a luxury. A GFCI is a simple device reasonably priced. If you are unsure about installation, seek a qualified electrician.

GFCIs protect against shock or electrocution when a plugged-in electrical product is dropped into a sink, pool, pond, puddle, or hot tub (a shock may be felt in the split second before the GFCI trips). A GFCI also cuts off current when a person contacts a product like an electric heater or an electric power tool, which may be "leaking electricity."

The National Electrical Code now requires GFCIs for protection in the bathroom, garage, kitchen and outdoor outlets of new homes.

Outlet type GFCI

Circuit breaker type GFCI

Portable type GFCI

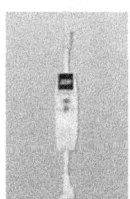

No GFCIs installed?

Buy one. GFCIs come in several models, including a portable plug-in type.

Attach a portable GFCI between the power receptacle and the plug of any electric saw, lawn edger, weed trimmer or other outdoor (or indoor) electrical equipment; or have a qualified electrician install receptacle or circuit breaker GFCI protection for your family. Make sure you have GFCIs for swimming pool underwater lighting circuits, for electric circuits of hot tubs, and for wall outlets within 20 feet of such pools as required in the National Electrical Code.

Testing GFCIs

To be sure your life-protecting GFCIs are working properly, use this test or the instructions that come with the GFCI.

1. Plug a night light (or radio turned up loud, if you have a circuit breaker GFCI) into a GFCI-protected wall outlet, and turn it on.
2. Press the GFCI test button or switch. The light or radio should go off.
3. Press the reset feature to restore power.

If the light or radio does not go off when the test button or switch is pressed, the GFCI is not working or is not wired correctly. Contact a qualified electrician to correct the problem or install a new GFCI.

A 3-pronged plug used in a 3-hole outlet protects against shock from a defective electrical product, cord or plug with grounding problems.

Electricity to power your electrical products travels along a path called a circuit. As long as it stays in its intended path while traveling to "ground," it does its job with minimal risk of electric shock. But when a product, cord or plug is damaged, out-of-path electricity may energize expose metal parts as it seeks a new path to ground. If you come in contact with energized conductive parts and provide a path to ground, the electricity will deliver a shock. The third prong on a plug is there to carry any stray electricity to ground through a 3-pronged receptacle.

Many electrical products designed for outdoor use have 3-pronged plugs (except for power tools and other products which may protect you against shock with a system of double insulation).

Never, ever, remove the third prong of a *3-prong plug*.

If your outdoor wall outlet has room for only 2 prongs, you should replace it with a GFCI-protected, 3-hole grounding type receptacle. When using a 3-to-2 grounding adapter, be certain that the receptacle itself is grounded or GFCI-protected for the adapter to work. Use a circuit tester (available in hardware stores) to find out if your outdoor receptacles are grounded, or call a qualified electrician to help you make sure.

Arc-fault circuit interrupters (AFCIs)

Problems in home wiring, like arcing and sparking, are associated with more than 40,000 home fires each year. These fires claim over 350 lives and injure 1,400 victims annually.

A new electrical safety device for homes, called an arc fault circuit interrupter or AFCI, is expected to provide enhanced protecting from fires resulting from these unsafe home wiring conditions.

Typical household fuses and circuit breakers do not respond to early arcing and sparking conditions in home wiring. By the time a fuse or circuit breaker opens to defuse these conditions, a fire may already have begun.

Requiring *AFCIs*

AFCIs are already recognized for their effectiveness in preventing fires. The most recent edition of the National Electrical Code, the widely-adopted model code for electrical wiring, requires AFCIs for bedroom circuits in new residential construction, effective January 2002.

Future editions of the code, which is updated every three years, could expand coverage to other circuits, including outdoor circuits.

Hazards in Use of Electrical Products Outdoors

How to Avoid Outdoor Electrical Accidents

Water	Keep outlets covered. Use a Ground fault circuit interrupter (GFCI). Keep products with line cords away from sinks, puddles, pools, ponds, and hot tubs. Keep outdoor outlets weather-protected with outlet covers.
Disabled 3-prong plug	Never remove third prong. Dispose of electrical items and extension cords with damaged prongs.
Damaged product wiring	Replace or have damaged parts, cords, plugs repaired by qualified professionals before use.
Improper product operation. Exposed blades or moving parts	Read instruction manual. Use goggles or other safety aides. Never bypass a safety device.
Unattended products	Switch off, unplug, store and lock products not in use.
Extension cord misuse	Match product power needs (on product labels and in manuals) to extension cord label information and make sure they are rated appropriately for outdoor use.

Improper product storage	Store outdoor electrical products indoors.
Overloaded branch circuits	Limit power use on each branch circuit to its rated capacity.
Use of indoor product outdoors	Use only weather-resistant products outdoors.
Power line contact	Contact your regional utility protection center (such as Digsafe, Call Before You Dig, or Miss Utility) to locate buried power lines before digging or drilling. Locate overhead power lines before trimming trees, flying kites or house painting, and keep ladders away.
Pad-mounted electrical equipment	Keep off and away from this electrical equipment. If you notice the cabinet doors or locks have been tampered with or left open, contact your local utility immediately.
Gasoline, naphtha fumes	Avoid where electrical sparks may cause fire or explosion.

For more information, contact CPSC (1-800-638-2772 or www.cpsc.gov), ESFI (703-841-3296 or www.electrical-safety.org) or your local utility company.

Before you use any electrical products outdoors:

✓ Make sure it was intended for outdoor use. Does the product's instruction manual or an attached label warn, "Not for Outdoor Use" or "Indoor Use Only"? Unless an electrical product is designed to be weather resistant, a sudden summer shower can ruin the product *and* turn it into a serious shock hazard. Most electrical products intended for continuous outdoor use have heavily insulated cords and molded-on plugs to prevent moisture from seeping in.

✓ Study all instructions carefully. Keep the instruction manual where you can easily find it. Reread it from time to time to refresh your memory.

✓ Inspect products for damaged cords, plugs or wiring. Turn the product off and unplug it if a cord overheats. Take a damaged product to the manufacturer's authorized repair center or have a qualified electrician repair it.

✓ Make sure a recognized testing laboratory certifies the product. This insures that the product is designed and manufactured in accordance with established safety standards. Look for these and other markings of internationally recognized testing laboratories:

SAFETY RULES

Follow these safety rules for every electrical product you use outdoors:

Outdoor portable electrical appliances and power tools should always be:

✓ Plugged in and turned on only when in use.

✓ Turned off and in lock position when being carried or hooked up to attachments like mower baskets or saw blades.

✓ Stored indoors (with a few exceptions such as electric barbecue grills, which can be covered to remain outdoors) and away from water and excessive heat.

✓ Used only when all safety guards are in place. Sharp blades and rapidly moving parts can cut off a finger or a toe.

Outdoor portable electrical appliances and power tools should never be:

✓ Left unattended outdoors, even when you leave temporarily. If there is a key, remove it. Put the product where no curious child or unqualified adult can misuse it.

✓ Plugged in while the switch is in the "on" position or while being carried or moved.

✓ Carried by their cords.

✓ Used while wet or close to water.

✓ Used near sharp edges or in conditions which can damage the product, its cord or its plug. Loose and broken wires are both shock and fire hazards.

✓ Repaired by anyone who is not a licensed electrician, authorized by the manufacturer or trained to repair the particular product.

Follow these rules to avoid water hazards:

✓ Keep outdoor outlets covered and dry between uses. New outlet covers are available that offer weather protection while a plug is inserted into the outlet.

✓ Except for electric snow blowers and other appliances designed for use in a wet environment, select a dry day to power-up outdoors.

✓ Keep cords and plugs away from sweating pipes and puddles.

✓ If an electrical product falls into water, make sure you are dry and not in contact with water or metal surfaces and unplug it immediately. Do not reach into the water for it.

✓ Use a ground-fault circuit interrupter (GFCI).

Photography by Mark Regan

HOT TUBS, SPAS, AND POOLS

Follow these rules to avoid hot tub, spa, and pool hazards:

- ✓ Keep outlets near hot tubs, spas and pools covered and dry between uses. New outlet covers are available that offer weather protection while a plug is inserted into the outlet.

- ✓ Keep cords and plugs away from hot tubs, spas and pools and puddles from wet bathers. Never handle electrical items, plugs or outlets when wet.

- ✓ If an electrical product falls into water, do not reach into the water for it. Make sure you are dry and not in contact with water or metal surfaces and unplug it immediately or shut off the circuit powering the item.

- ✓ Hot tubs, spas and pools, and outlets on or near them should be protected by a ground-fault circuit interrupter (GFCI). Many older swimming pools that pre-date the introduction of GFCIs in the 1970s should be upgraded to add GFCI protection for branch circuits supplying power to underwater pool lights operating above 15 volts, and outlets within 20 feet of the pool.

Note, however, that when a person is immersed in an isolated body of water, like a hot tub, the water could become electrified without involving a ground fault as the electric current passes through water (and perhaps a person) from one electrical pole to the opposite pole. In this case, the GFCI may not provide shock or electrocution protection.

EXTENSION CORDS

Guidelines for selecting and using outdoor extension cords:

- ✓ Use only extension cords marked "For Outdoor Use." Weather-resistant, medium-to-heavy gauge extension cords have connectors molded onto them to prevent moisture from seeping in and outer coatings that are designed to withstand being dragged along the ground.

- ✓ Outdoor extension cords come in 25 to 150 foot lengths. Buy only the length you need. Above 100 feet you can lose power—a hazard when using power tools.

- ✓ Use three-wire extension cords with 3-pronged plugs. Exception: Extension cords for use with appliances and tools that are "double-insulated."

- ✓ Completely connect plugs. Push them in all the way. *Do not plug one extension cord into another.*

- ✓ Unwind cord before using. Do not use if damaged. Do not cover or walk on cords.

✓ Never leave an open line (no product plugged into the end of an extension cord while it is plugged into an outlet). Not even for a minute. Always unplug cords not in use.

✓ Never leave extension cords outside in the snow or very cold weather for extended periods.

✓ Replace outdoors extension cords every three or four years if damage is noted.

Match each outdoor electrical product to its extension cord:

✓ Match power needs (amperage) of electrical products with amperage rating of extension cords.

✓ The extension cord capacity should be as high as or higher than that of the electrical product attached to it. Amperage ratings for outdoor electrical products can range from "1 A" for a bug killer to "15 A" for a snow blower and are found on nameplates attached to products. Compare them to the rating information on extension cord packaging and on labels permanently attached to cords.

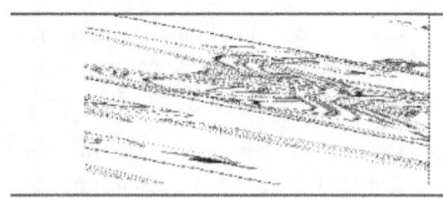

> To convert amps to watts, multiply by 120 volts.
> For example, 10 A x 120 V = 1200 W.

Match the extension cord gauge to the amperage rating of the product. AWG on the above label stands for American Wire Gauge. Cords for outdoor use are generally either 12 AWG (heavy) or 14 AWG (medium).

ELECTRICAL LAWN & GARDEN PRODUCTS

Follow every general safety rule for outdoor electrical products when using electrical lawn and garden products. Then take some extra precautions.

Lawnmowers and other lawn and garden equipment with sharp blades and rapidly moving parts can cause serious injury by cutting off a finger or a toe. Never remove the guards.

Keep children well away from lawnmowers and other products, which can throw objects such as rocks and sticks.

Products like power shovels or diggers, lawnmowers, mulchers, tillers, thatchers and leaf or snow blowers move and have moving parts that can cut, burn, even blind when directions are not followed. Study each product's manual for safe operation rules.

Photography by Mark Regan

Mowing a lawn:

- ✓ Clean area first; remove rocks, branches, wires, bones or other foreign objects that can be thrown by blades.
- ✓ Avoid wet grass. Mow only in daylight.
- ✓ Always wear enclosed shoes.
- ✓ Never remove safety guards or adjust wheel height while motor is running.
- ✓ Keep cords out of working path.
- ✓ Avoid loose clothing and jewelry that can catch on moving parts.
- ✓ Keep bystanders, especially children away.
- ✓ Push, don't pull. Mow across not up and down slopes.
- ✓ Clipping, trimming a hedge or edging, wear safety goggles or other protection recommended by the manufacturer. Never overreach especially when on a ladder.
- ✓ Avoid power lines. Contact can cause serious injury or death.

Use these accident-prevention techniques:

- ✓ Keep your equipment in good operating condition.
- ✓ Blocked snow or leaf blower: Unplug the power cord for these electric appliances (or

turn the engine off for gasoline-powered products) before attempting to clear the obstruction. To clear out the blockage, use a stick long enough to protect your hands from injury. Never put your hand near the collection or discharge chutes. Even when the engine is off, blades can remain spring-charged, resulting in swift movement when its path is cleared. **Extra precaution is always wise.**

Ladders and electricity do not mix.

Electrocutions can occur when ladders are used near overhead wires to clean gutters, paint houses, trim trees and repair roofs and chimneys or install outdoor antennas.

- ✓ Use only a fiberglass or wooden ladder if you must work near overhead wires and do not let it come into contact with the wires.

- ✓ If you must use a metal ladder, keep it well away from overhead lines.

- ✓ If a ladder starts to fall into an overhead line, let it go! Stay nearby while someone else calls the power company to cut off electricity to the line before you touch or move the ladder that is in contact with a power line.

- ✓ Never touch a person who is holding a ladder that has fallen onto a power line. Use something that does not conduct electricity, such as a long piece of dry wood or rope to push or pull them loose.

BATTERY-OPERATED PRODUCTS

Follow the same safety rules with cordless, battery-operated products as any other electrical product. Batteries generate electric power. Read and follow manufacturer's instructions.

Some special things to remember when using battery-powered products:

✓ Keep batteries away from children.

✓ Cordless products, since they don't have to be plugged in, are always ready to use. Store them away from children or inexperienced persons.

✓ Bring cordless products indoors overnight so they won't be subjected to a higher moisture level or a sudden rainstorm.

✓ Remove batteries or lock switches in "off" position when not in use before changing accessories or cleaning battery-operated products to prevent accidents while your hands are near blades or other moving parts.

- ✓ All batteries should be replaced at the same time. Do not mix fresh and discharged batteries or battery types.

- ✓ Ensure batteries are installed correctly in device and charger with regard to polarity (+ and -).

- ✓ Do not use cordless tools near gaseous or explosive materials. Sparks from their motors might cause fires or explosions.

- ✓ Never short circuit batteries as this may lead to high temperatures, leakage or explosion.

- ✓ Never attempt to disassemble batteries as this can lead to electrolyte burns.

Things to remember when recharging batteries:

- ✓ Always recharge battery-operated products with the charging unit and procedure recommended by the manufacturer.

- ✓ Recharge products in a dry place away from radiators, heaters, stoves, flames or chemicals.

- ✓ Plug charger directly into an electrical outlet, never into an extension cord.

- ✓ If your product battery does not recharge properly, first check the trouble section of your instruction manual. Next, take the product and the charger to a manufacturer-recommended repair center.

✓ Replace batteries only with recommended size and type to insure compatibility between rechargeable battery and charging circuit.

✓ Never attempt to recharge primary batteries as this can cause them to leak, cause a fire or explode.

Take these precautions with extra batteries:

✓ Do not expose batteries to moisture, frost or temperatures over 110 degrees or under 20 degrees F. Do not store in refrigerator or freezer. If batteries get cold; bring them to room temperature before use.

✓ Do not store batteries touching metal objects such as wire, nails or coins (in your pocket). Such contact can cause a large current flow, possibly leading to burns or fire.

And for safe battery disposal:

Batteries and battery packs can explode in a fire. Follow manufacturer's instructions for disposal.

POWER TOOL SAFETY

Power tools are often used out of doors or in a garage or shed where the door should be open for adequate ventilation. Power tools require skilled use. Operators should read the product instruction manual.

Power tools should never be used when children are in, or even near, the work area.

Power tools should always be:

✓ Held by the insulated gripping surface to avoid electrical shock.

✓ Used with safety goggles and other safety gear: a face shield, dust mask, hard hat, ear protection, gloves or safety shoes as recommended by the manufacturer.

✓ Used with a GFCI, either permanently installed or a plug-in type.

✓ Plugged into a three-pronged outlet known to be grounded, unless they are double insulated.

✓ Used with a three-wired extension cord, if needed.

✓ Used in a dry area away from explosive fumes (gasoline or naphtha), dust or flammable materials.

Power tools should never be:

✓ Used while wearing loose clothing or jewelry that can get caught in a moving part.

✓ Used near live electrical wires or water pipes, especially when cutting or drilling into walls where they could be accidentally touched or penetrated.

✓ Used after they have tripped a safety device such as a GFCI. Take the tool to a manufacturer-authorized repair center for service.

✓ Used without guards or with an extension cord longer than 100 feet.

Other outdoor electrical products such as fans, bug killers, holiday or party lights, heaters, music systems, power paint rollers, barbecue spits and many more each have manufacturer-recommended precautions included in the instructions that are packaged with them. Take time to read and follow instructions. Here are a few reminders:

Power washer—This product uses water with electricity. Make sure you read the directions carefully.

Barbecue grill—Read directions to find out if it can be stored outdoors or used on an apartment balcony, patio or deck. Also check with your apartment building manager for usage rules and/or local ordinances or regulations.

Charcoal igniter—Do not store outdoors.

POWER LINE SAFETY

Before you work or move equipment around any power lines, you should know that

- ✓ Power lines kill more workers than any other electrical source—an average of 133 per year.
- ✓ Power lines are not insulated for contact.
- ✓ You should stay at least 10 feet away from power lines.
- ✓ You can be electrocuted by a power line, even if you are wearing rubber gloved and rubber-soled boots.

Electrical accidents rank sixth among all causes of work related deaths. One worker is killed by electricity nearly every day, and power lines kill more workers than any other electrical hazard.*

Workers carrying ladders, crane operators, boom truck operators and those doing yard work near power lines should stay clear of the lines. Those digging near underground lines should check with local utilities and know to avoid utility hazards. For information on downed power lines, see page 34.

For more information on power line safety, visit these following web sites:
> www.osha.gov
> www.nfpa.org
> www. cdc.gov (use their search feature to find information on "Construction Occupational Safety and Health")
> www.buildsafe.org
> www.nsc.org

*Data from "Occupational Injuries in the U.S., 1992–1998," published in the *Journal of Safety Research* 34 (2003), pp. 241–248.

ELECTRICAL SAFETY PRECAUTIONS DURING STORMS

The Electrical Safety Foundation International (ESFI) warns consumers to beware of the dangers hurricanes and floods cause when water comes in contact with electricity.

ESFI offers this safety advice:

Flooded Areas—Take care when stepping into a flooded area, and be aware that submerged outlets or electrical cords may energize the water, posing a potential lethal trap.

Wet Electrical Equipment—Do not use electrical appliances that have been wet. Water can damage the motors in electrical appliances, such as furnaces, freezers, refrigerators, washing machines, and dryers. A qualified service repair dealer should recondition electrical equipment that has been wet.

Portable Generators—Take special care with portable electric generators, which can provide a good source of power, but if improperly installed or operated, can become deadly.

Do not connect generators directly to household wiring. Power from generators can backfeed along power lines and electrocute anyone coming in contact with them, including lineworkers making repairs. A qualified, licensed electrician should install your generator to ensure that it meets local electrical codes. Other tips include:

✓ Make sure your generator is properly grounded.

- ✓ Keep the generator dry.
- ✓ Plug appliances directly into the generator.
- ✓ Make sure extension cords used with generators are rated for the load, and are free of cuts, worn insulation, and have three-pronged plugs.
- ✓ Do not overload the generator.
- ✓ Do not operate the generator in enclosed or partially enclosed spaces. Generators can produce high levels of carbon dioxide very quickly, which can be deadly.
- ✓ Use a ground fault circuit interrupter (GFCI) to help prevent electrocutions and electrical shock injuries. Portable GFCIs require no tools to install and are available at prices ranging from $12 to $30.

Downed Power Lines—These can carry an electric current strong enough to cause serious injury or possibly death. The following tips can help you stay safe around downed lines:

- ✓ If you see a downed power line, move away from the line and anything touching it. The human body is a ready conductor of electricity.
- ✓ The proper way to move away from the line is to shuffle away with small steps, keeping your feet together and on the ground at all times. This will minimize the potential for a strong electric shock. Electricity wants to

- ✓ move from a high voltage zone to a low voltage zone—and it could do that through your body.

- ✓ If you see someone who is in direct or indirect contact with the downed line, do not touch the person. You could become the next victim. Call 911 instead.

- ✓ Do not attempt to move a downed power line or anything in contact with the line by using another object such as a broom or stick. Even non-conductive materials like wood or cloth, if slightly wet, can conduct electricity and then electrocute you.

- ✓ Be careful not to put your feet near water where a downed power line is located.

- ✓ If you are in your car and it is in contact with the downed line, stay in your car. Honk your horn for help and tell others to stay away from your vehicle.

- ✓ If you must leave your car because it's on fire, jump out of the vehicle with both feet together and avoid contact with the live car and the ground at the same time. This way you avoid being the path of electricity from the car to the earth. Shuffle away from the car.

- ✓ Do not drive over downed lines.

LIGHTNING SAFETY

Lightning strikes the United States as many as 20 million times each year. Because lightning traditionally causes more deaths than tornadoes or hurricanes and occurs when outdoor activity reaches a peak, those who work outdoors should be aware of lightning safety guidelines.

Outdoors is the most dangerous place to be during a lightning storm. Because lightning can travel sideways for up to 10 miles, blue skies are not a sign of safety. If you hear thunder, take cover.

- ✓ If outdoors, go inside. Look for a shelter equipped with a lightning protection system.

- ✓ Go to a low point. Lightning hits the tallest object. Get down if you are in an exposed area.

- ✓ Stay away from trees.

- ✓ Avoid metal. Don't hold metal items, including bats, golf clubs, fishing rods, tennis rackets or tools. Avoid clotheslines, poles and fences.

- ✓ If you feel a tingling sensation or your hair stands on end, lightning may be about to strike. Crouch down and cover your ears.

- ✓ Stay away from water. This includes pools, lakes, puddles and anything damp, such as wet poles or grass.

- ✓ Don't stand close to other people. Spread out.
- ✓ Once indoors, stay away from windows and doors.
- ✓ Do not use corded telephones except for emergencies.
- ✓ Unplug electronic equipment before the storm arrives and avoid contact with electrical equipment or cords during storms.
- ✓ Avoid contact with plumbing, including sinks, baths and faucets. Do not take baths and showers during electrical storms.
- ✓ Don't forget pets during thunderstorms. Doghouses are not lightning-safe. Dogs that are chained can easily fall victim to a lightning strike.

Data from the National Weather Service shows that lightning strikes are fatal in approximately 10 percent of strike victims. Another 70 percent of survivors suffer serious long-term effects. Victims of lightning strikes should be given CPR if necessary, and seek medical attention.

GLOSSARY

Amperage (amps)—A measure of electrical current flow.

Arc-fault circuit interrupter (AFCI)—Protection from fires caused by affects of electrical arcing in wiring. AFCI device will de-energize the circuit when an arc fault is detected.

Circuit breaker or fuses—Protect against over-current and short circuit conditions that could result in potential fire hazards and explosion.

Electrical faults—A partial or total failure in an electrical conductor or appliance.

Energized—Electrically connected to a source of potential difference, or electrically charged so as to have a potential different from that of the ground.

Gauge—Standard or scale of measure.

Ground-fault circuit interrupter (GFCI)—Protection against shock and electrocution. GFCI device will de-energize a circuit when it senses a difference in the amount of electricity passing through the device and returning through the device, or a "leak" of current from the circuit.

Grounded/grounding—A conducting connection, whether intentional or accidental, by which an electric circuit or equipment is connected to the earth, or to some conducting body of relatively large extent that serves in place of the earth.

Overcurrent—Any current in excess of the rated current or ampacity of a conductor. May result in risk of fire or shock from insulation damaged from heat generated by overcurrent condition.

Outlet—A contact device installed along a circuit for the connection of an attachment plug and flexible cord to supply power to portable equipment and electrical appliances. Also known as receptacles.

Three-pronged plugs and outlets—Protect against potential shock from the use of damaged products or electrical power cords designed to take stray electrical current safely to ground.

Short circuits—An abnormal electrical path.

Voltage (volts)—A measure of electrical potential.

Wattage (watts)—A measure of the rate of energy consumption by an electrical device when it is in operation.

About the Electrical Safety Foundation International

The Electrical Safety Foundation International is a not-for-profit (501)(c)(3) organization whose board of directors and officers serve without compensation.

Board of Directors

Ball State University
Connector Manufacturing Company
Cooper Power Systems Division
CSA Group
DuPont
Eaton/Cutler-Hammer
General Cable
Int'l Association of Electrical Inspectors
Int'l Brotherhood of Electrical Workers
Intertek-ETL Semko Americas
Leviton Manufacturing Company Inc.
Mr. Electric
National Consumers League
National Electrical Contractors Association
National Electrical Manufacturers Association
National Fire Protection Association
Pass & Seymour/Legrand
Siemens Energy & Automation
Underwriters Laboratories Inc.
USDA Extension Service

Beacon Endowment Contributors

The ESFI Board established the ESFI Safety Awareness Fund through the "Light a Beacon for Safety" Campaign to secure financial support to implement new electrical safety education and awareness initiatives. We salute the following companies for their leadership and support:

Platinum Beacon Sponsors ($500,000 or greater)
Square D Company/Scheider Electric

Gold Beacon Sponsors ($250,000–$499,000)
Cooper Industries, Inc.
Eaton/Cutler-Hammer
GE
GE Consumer Products
GE Medical Systems
Rockwell

Silver Beacon Sponsors ($100,000–$249,999)
ABB Power T & D Company, Inc.
Advance Transformer Company
CSA International
Emerson
Hubbell Incorporated
Intertek Testing Services
Leviton Manufacturing Company Inc.
Lithonia Lighting
NECA-IBEW NLMCC
Siemens Corporation
Thomas & Betts Corporation
Underwriters Laboratories Inc.

Bronze Beacon Sponsors ($50,000–$99,999)
Edison Electric Institute
General Cable Corporation
Graybar Electric Company Foundation

Legrand North America
   Pass & Seymour/Legrand
   The Watt Stopper
   Wiremold
Lutron Electronics Company, Inc.

Copper Beacon Sponsors ($25,000–$49,999)

Connector Manufacturing Company
FCI USA, Inc.
Genlyte Thomas Group, LLC
   Lightolier
ILSCO Corporation
Phelps Dodge Foundation
Phoenix Contact, Inc.
Robroy Industries, Inc.
S & C Electric Company
Southwire Company
The Electrical Contracting Foundation

Beacon Contributor (up to $24,999)

Advanced Protection Technologies
Alcan Cable
American Lighting Association
L3 Communication
Lamson & Sessions
MGE UPS Systems, Inc.
National Association of Electrical
   Distributors

2004–2005 Annual Contributors

(as of 4th quarter, 2005)

BENEFACTORS ($25,000 and greater)
CSA International
Square D Company/Schneider Electric
Underwriters Laboratories Inc.

PATRONS ($10,000–$24,999)
GE
 GE Medical Systems
 GE Consumer Products
Hubbell Incorporated

SPONSORS ($5,000–$9,999)
Advance Transformer
Cadet Manufacturing
Cooper Industries, Inc.
Copper Development Association
Eaton/Cutler-Hammer
Edison Electric Institute
FirstEnergy Foundation
Hypertherm Incorporated
Intertek Testing Services
Legrand North America
 Pass & Seymour/Legrand
 The Watt Stopper
 Wiremold
Leviton Manufacturing Company Inc.
Lincoln Electric
Lutron Electronics Company, Inc.
Mr. Electric
National Electrical Contractors Association
National Fire Protection Association
Panasonic Corporation of North America
Siemens Energy & Automation

CONTRIBUTORS (up to $4,999)
Ameren Services
American Lighting Association
 Education Foundation
American Public Power Association

Cantex Inc.
Cleco Corporation
Coleman Cable Systems, Inc.
Connector Manufacturing Company
FERRAZ-SHAWMUT, Inc.
Hoffman Enclosures Inc.
IBEW #51
IBEW #252
IBEW #613
IBEW #1049
ILSCO Corporation
International Brotherhood of Electrical Workers
LMCC of Chicago
   IBEW Local #134
   Electrical Contractors Association (ECA)
Nat'l Electrical Manufacturers Representatives Association (NEMRA)
National Rural Electric Cooperative Association
NECA IBEW Local 176
NECA IBEW Local 701 Labor Management Cooperation Committee
   Electrical Contractors Northeast Illinois
   NECA Northeast Illinois Chapter
Northern Indiana Public Service Company
OSRAM Sylvania
PEPCO
Phoenix Contact Inc.
PPL Services Corporation
Radix Wire Company
Regal-Beloit Corporation
S&C Electric Company
S&S Electric
Southwire Company
The Homac Companies
Warren LMCC
WESCO Distribution, Inc.
Youngstown Area LMCC

www.ingramcontent.com/pod-product-compliance
Lightning Source LLC
Chambersburg PA
CBHW071542170526
45166CB00004B/1514